Pre-Conflict Management Tools: Winning the Peace

Aaron B. Frank

Center for Technology and National Security Policy
National Defense University

February 2005

The views expressed in this article are those of the author and do not reflect the official policy or position of the National Defense University, the Department of Defense, or the U.S. Government. All information and sources for this paper were drawn from unclassified materials.

Aaron B. Frank is an Associate at Booz Allen Hamilton. Mr. Frank was a Research Associate in the Center for Technology and National Security Policy from 2001 through 2004, where he supported the Program in Computational Social Science Modeling and the Pre-Conflict Management Tools Program.

Defense & Technology Papers *are published by the National Defense University Center for Technology and National Security Policy, Fort Lesley J. McNair, Washington, DC. CTNSP publications are available online at* http://www.ndu.edu/ctnsp/publications.html.

Contents

Executive Summary

The Pre-Conflict Management Tools (PCMT) Program was developed to transform how intelligence analysts, policy analysts, operational planners, and decisionmakers interact when confronting highly complex strategic problems. The PCMT Program capitalizes on technologies and methods that help users collect, process, perform analyses with large quantities of data, and employ computational modeling and simulation methods to determine the probability and likelihood of state failure. The Program's computational decision aids and planning methodology help policymakers and military planners devise activities that can mitigate the consequences of civil war, or prevent state failure altogether.

State failure has become an increasingly important national and international security issue since the end of the Cold War. Weak and failed states establish a nexus of interests between global terrorism, embattled leaders or insurgents, and large populations easily mobilized by a combination of violent ideology and economic opportunity. Civil war, the most common form of armed conflict around the world, undermines regional and international stability and catalyzes larger national security problems, such as weapons proliferation, organized crime, and terrorism.

The PCMT Program builds on social science research on state failure and conflict, by turning government users into consumers of social science models employed by academic researchers and validated through peer review processes and implementation by practitioners. By constructing an analytic suite out of existing models, the Program avoids the controversies of 1960's social science research programs, such as Project Camelot, by rejecting the notion of a single, government-sponsored theory of conflict or placing policymakers in the position of determining what is or is not valid social science.

PCMT architecture and methodology capitalize on changes in the landscape of information made possible by the ever-increasing quantity and diversity of information available electronically, by modeling, simulation, and analysis for identifying social vulnerabilities, and by a collaborative analytic and planning process at interagency and international levels. Each component of the PCMT architecture incorporates or extends established tools and practices that have improved performance in a variety of endeavors in other domains.

The PCMT data collection capability helps the user organize and exploit all information available in electronic form, whether collected from open sources or the user's private databases. This enables analysts to filter data, rather than sample from small populations of sources whose ability to represent the character of the available universe of data is in doubt. Moreover, automated document coding enables analysts to work from datasets that would be too costly to construct, maintain, and manipulate manually. As a result, PCMT data collection and management technologies enable users to perform new kinds of analysis.

The PCMT modeling and simulation suite contains multiple models of social vulnerabilities that assess the probability of state failure. Each model instantiates a different social science theory as to why states fail and civil war occurs. The suite gives policymakers diverse perspectives on each country or region. The application of multiple, competing models in analytic processes also assists users in confronting uncertainty by preventing decisionmakers from developing plans based on the outputs of a single model or theory. Instead, the PCMT suite assists users in crafting robust, adaptive policies that satisfice across landscapes of potential futures or scenarios generated through simulation. PCMT also gives users an understanding of network dynamics within a region, which enables identification of regional elites and relationships that can help or hinder the achievement of U.S. political objectives.

Finally, PCMT is constructed to facilitate communication and analysis at interagency and international levels. By giving users warning of state failure months in advance, coalitions, partnerships, and plans can be formed to head off a crisis. However, such actions demand coordination between multiple agencies and the resources of local and international groups. They also require an analytic infrastructure that accounts for multiple, competing models of political and social behavior and the balancing of political objectives in order to devise strategies that achieve a balance between near-term and long-term objectives, and political, economic, military, and social goals.

The PCMT Program demonstrated the potential of computational tools and social science models to aid the development of robust, adaptive policies. Transitioning PCMT from a research program into an operational program will provide a new and necessary capability for winning wars and, more importantly, the peace.

Introduction

In April 2003, the Center for Technology and National Security Policy (CTNSP) at National Defense University (NDU) launched a research project designed to give the U.S. government new tools for preventing the outbreak of violent conflict. The Pre-Conflict Management Tools (PCMT) project is the culmination of several years of working with Regional Combatant Commanders (RCCs) on identifying gaps between military capabilities and strategic requirements. RCCs repeatedly noted that they had numerous means for employing military force, but little capability for preventing the outbreak of violent conflict.[1] Their message served as a constant reminder of the difference between winning the war and winning the peace and of the tensions between near-term and long-term outcomes and stability.

PCMT seeks to provide users with a broad base of knowledge, perspectives, and insights from the social sciences. The formal incorporation of social science theories and models into the defense community is an important step in developing capabilities to confront transnational threats such as civil war, terrorism, international crime, environmental catastrophes, and infectious disease. While these threats may be low-level, or build slowly, the ability of single states or government agencies to deal with them once manifest is poor. Policymakers require tools that identify problems early if they are to be successful in mobilizing resources and reaching a consensus at interagency and international levels regarding the appropriate actions to take to prevent the emergence of a full-scale crisis. Doing so, however, demands that a breadth of social science models be employed as analytic and decision aids to make sense of large volumes of ambiguous, contradictory, but relevant data.

After an 18-month proof of concept study, the PCMT program identified real-world operational requirements and demands, and demonstrated the ability to fill in existing gaps between requirements and capabilities using social science models. One reason for the program's success is that it integrates existing technologies, many of which are already employed by the Department of Defense and elsewhere, into a synergistic whole. A second reason is that the program has built upon RCC experiences and lessons learned from a variety of international crises, military operations, and diplomatic initiatives. PCMT is a collaborative effort that includes contributions from the Defense Advanced Research Projects Agency (DARPA), the U.S. Army Intelligence and Security Command (INSCOM), and Joint Forces Command (JFCOM). It is focused on conflict prevention in the Central Asian states of Kazakhstan, Kyrgyzstan, Tajikistan, Turkmenistan, Uzbekistan, Afghanistan, and Pakistan. Once deployed, PCMT will provide users with new capabilities for shaping the international system and mitigating conflicts before they occur. Integrating components for data collection, situational and policy analysis, and planning will provide PCMT users a powerful suite of tools and transform how policymakers and operators understand, confront, and shape the international system.

---

[1] Discussions with PCMT Program Manager Joseph Eash and Deputy Program Manager Desmond Saunders-Newton regarding their experience managing the Advanced Concept and Technology Demonstration programs.

The first half of this paper provides readers a context for why the PCMT program is an important development and discusses the complexities involved in incorporating social science models into the RCC and interagency planning processes. The second half provides an overview of the structure, methods, and technologies of PCMT and outlines a future research agenda for making it deployable to RCCs and interagency and coalition planners.

Meeting the Demands of Transformation

For many, social science theory and models have little significance in the larger pursuit of military transformation. Indeed, the Pentagon's major transformation initiatives focus almost exclusively on the consequences of technological change and the requirement for new weapons, organizations, and operational concepts. Despite the lack of an overt relationship between the social sciences and military transformation, understanding and influencing human behavior, both individually and collectively, remain the ultimate objectives of national security institutions.

For almost three decades, military planners and strategists have been concerned with the future of warfare, particularly changes resulting from technological innovation. Over this period, concepts have evolved from the Soviet inspired Military Technical Revolution (MTR), to the Revolution in Military Affairs (RMA), to the resulting collection of concepts and vision categorized as military transformation. What began as a primarily weapons oriented approach to understanding military change in the 1960s and 70s–MTR–grew to include doctrine and organization under the term RMA.[2] Later, the introduction of the term "transformation" signified the incorporation of an even broader set of concerns and ideas, including the relationship between the military and society, military capabilities and the structure of the international system, and the emergence of new political and military actors. Military transformation has gone beyond the narrow emphasis on technological and operational change and embraced broader social and political aspects, such as the coupling between the structure of the international system and warfare, often referred to as military revolution theory.[3] As the context surrounding innovation and change has become increasingly aware of–and responsive to–political and

---

[2] The Soviet concept of the Military Technological Revolution was first applied to the implications of nuclear weapons for warfare and strategy. This concept was later expanded to include the development of advanced conventional strike capabilities, command and control systems, sensors and intelligence, and computer processing. While the term Revolution in Military Affairs is normally credited to U.S. strategic planners, the term was in use by Soviet military planners in the 1970s. See John Erickson, Edward L. Crowley, and Nikolai Galay, eds., *The Military-Technical Revolution* (New York, NY: Praeger, 1966); Gen. N.A. Lomov, ed., "Scientific-Technical Progress and the Revolution in Military Affairs (A Soviet View)," *Soviet Military Thought*, No. 3 (Washington, D.C.: United States Air Force, 1973), United States Air Force, trans.; Eliot A. Cohen, "A Revolution in Warfare," *Foreign Affairs*, Vol. 75, No. 2 (March/April 1996), 39-41; and MacGregor Knox and Williamson Murray, "Thinking About Revolutions in Warfare," in MacGregor Knox and Williamson Murray, eds., *The Dynamics of Military Revolution, 1300-2050*, (Cambridge: Cambridge University Press, 2001) 2-4.

[3] See William H. McNeill, *The Pursuit of Power* (Chicago, IL: University of Chicago Press, 1982); Brian M. Downing, *The Military Revolution and Political Change* (Princeton, NJ: Princeton University Press, 1992); Clifford J. Rogers, ed., *The Military Revolution Debate* (Boulder, CO: Westview Press, 1995); Geoffrey Parker, *The Military Revolution: Military Innovation and the Rise of the West, 1500-1800* (New York, NY: Cambridge University Press, 1996); Michael Howard, *War in European History* (New York, NY: Oxford University Press, 2001); Jeremy Black, *War and the World: Military Power and the Fate of Continents, 1450-2000* (New Haven, CT: Yale University Press, 2001); and Philip Bobbitt, *The Shield of Achilles* (New York, NY: Anchor Books, 2002).

social conditions, transformation has become an imperative driven by decisionmakers' beliefs about opportunities and vulnerabilities.[4]

PCMT responds to this imperative by focusing on the detection of unstable political and social structures. Emphasizing the structure and shape of the international system changes the context for considering the use of force, which in turn generates demand for new technologies, organizations, and concepts. Comparing the roles and uses of military force during and after the Cold War reveals important change in the demands on military organizations.[5] Throughout the Cold War, the international system was largely viewed as stable; the primary role of U.S. military force was to deter the Soviet Union and its client states from taking actions to destabilize the balance of power.[6] By comparison, following the collapse of the Soviet Union, the U.S. military has been called on to push, nudge, and otherwise direct the international system toward a more a stable structure. The result has been the deemphasizing of deterrence and a greater reliance on engagement and interventions designed to encourage, persuade, compel, coerce, or otherwise influence foreign populations, both friendly and hostile, and arrest or contain dangerous process, such as economic or environmental crises.

The events of September 11, 2001 further reveal the significance of social factors for national security and military transformation. The Global War on Terrorism (GWOT), globalization, and the more general competition over political and social integration and fractionalization require the harmonization of all levers of national power to influence social and political conditions abroad. The application of military force, no matter how precise, cannot bring these conflicts to a peaceful conclusion. From this perspective, military transformation must therefore address the question of how to employ national capabilities, old and new, in an unfamiliar and dynamic environment. Information operations, precision-strike, and other capabilities cannot succeed if they are treated as little more than new ways to strike old targets—their application demands a new understanding of how the world works. Social science theories must be incorporated into the transformation agenda.

To understand the significance of social science theories and analysis to military operations and strategy it is useful to consider the case of strategic bombing and the development of airpower. The airplane was not sufficient to alter the nature of warfare. Aside from the operational demands for overhead imagery, radio navigation, and weather forecasting, a new understanding of social structures and processes, particularly political will and economic production were needed to make strategic bombing a reality. While the airplane provided the ability to bypass enemy defenders in favor of more lucrative targets, determining what those targets should be required new theories of how states were constructed, how political will was influenced through the use of force, and what

---

[4] See Douglas McGray, "The Marshall Plan," *Wired*, Vol. 11, No. 2 (February, 2003) available at http://www.wired.com/wired/archive/11.02/marshall.html.

[5] See Dana Priest, *The Mission* (New York, NY: W.W. Norton & Company, 2004), 21-40.

[6] On the desired stability of the bipolar, Cold War international system see Kenneth N. Waltz, *Theory of International Politics* (Reading, MA: McGraw Hill, 1979); and John J. Mearsheimer, "Why We Will Soon Miss the Cold War," *The Atlantic Monthly*, August 1990, available at http://www.theatlantic.com/politics/foreign/mearsh.htm.

the contribution of specific industries or facilities were to the economy and polity.[7] PCMT marks an important step in military transformation by enabling a better sense of the social and political effects of new weapons, concepts, and organizations.

---

[7] See David A. MacIssac, "Voices from the Central Blue: The Air Power Theorists," in Peter Paret, ed., *Makers of Modern Strategy* (Princeton, NJ: Princeton University Press, 1986), 624-647; John F. Kreis, ed., *Piercing the Fog* (Washington, D.C.: Air Force History and Museums Program, 1996); and Williamson Murray, "Strategic Bombing: The British, American, and German Experiences," in Williamson Murray and Allan Millett, eds., *Military Innovation in the Interwar Period* (New York, NY: Cambridge University Press, 1996), 96-143.

A New Approach

In developing PCMT, the program focused on political and social conditions as its measure of effectiveness. The result was that military operations were considered in a broad context, and other elements of national power were simultaneously considered. Moreover, analysis and planning sought to account for the near and long-term consequences of actions, creating a demand for new methods for considering effects and articulating and prioritizing political goals and objectives. Any credible effort to shape the international system would require interagency or international participation drawing upon the capabilities of a broad array of formal and informal organizations and participants with different cultures, perspectives, and interests. As a result, the PCMT analytic suite is designed to capture and explore a variety of perspectives that individually or collectively reflect the interests and concerns of different stakeholders. Moreover, PCMT architecture can be scaled and compartmentalized, enabling the inclusion of allies, international government organizations (IGOs), non-governmental organizations (NGOs), academics, corporations, and others who possess relevant knowledge, skills, or capabilities, but whose participation may be limited, highly specialized, or only known to a small group.

The creation of a common analytic environment and communications infrastructure establishes a collaborative environment but complicates policy formation because of the wide variety of views held by its participants, their differential access to, or trust in, datasets and sources, and beliefs about the efficacy of various actions at political and social levels. Traditional interagency planning and coordination result in the creation of an issue-oriented czar or oversight body that attempts to rationalize the activities of competing agencies and keep the policy and planning process moving by placing and keeping participants on the same page. However, strategies that emphasize hierarchy and appeals to authority demand the active involvement of senior executives and rarely produce intended results.[8]

Rather than push diverse contributors toward conformity, PCMT designers encourage diversity at informational, theoretical, and analytical levels. The result embraces competitive analysis, where competing models, assumptions, and data are used in parallel when conducting situational assessments, allowing PCMT to embrace competing world-views simultaneously. Whereas hierarchically driven interagency or international planning approaches are fragile and not highly scalable, the PCMT diversity-embracing, competitive process is highly scalable and can create robust policies that perform well across a range of models and datasets, rather than optimally given a singular set of data and analytic model. This marks a fundamentally different approach to the management of complex problems, where large numbers of heterogeneous actors must work together, or when deep uncertainty about the phenomenon in question exists.

---

[8] See James Q. Wilson, *Bureaucracy* (New York, NY: Basic Books, 2000), 264-274.

Civil War and Warning

Military transformation, interagency planning, and the competitive use of social science models could be applied to any number of problems in the international system. PCMT chose to address the problem of civil wars, because of their national, regional, and international effects, the requirement for international resources to prevent or contain them, and the difficulty of providing policymakers with timely warning of state failure.

The RCC demand for tools to prevent the outbreak of a conflict was a clear sign for the need to identify, confront, and reverse destabilizing trends, behaviors, and conditions before they fester and grow, i.e., the need for preventive defense.[9] From this perspective, civil wars and state failure are problematic, because they can restrict access to valuable resources or strategic geography, threaten the lives of American citizens within the county, provide terrorist or criminal groups with operational sanctuary, spread extremist or violent ideology, and destabilize neighboring states and entire regions.

PCMT conflict prevention efforts focus on discovering, arresting, and reversing processes that lead to large-scale political and social violence. Conflict prevention requires that sufficient warning be given to policy-makers, and that they devote the necessary time and resources to act on it. Given the prominence of civil war, insurgency, terrorism, and other conflicts that pit national governments against violent dissidents, PCMT analytic models have focused on identifying the conditions that give rise to state failure and civil war.

Civil wars have emerged as the most prevalent type of contemporary armed conflict, and they have long-term effects on regional security and prosperity. During the Cold War, the number of conflicts waged between nation-states dramatically declined, while the number of civil wars rapidly increased. Moreover, even in cases of interstate war, such as the GWOT, elements of state failure play a crucial role as violent, radical groups take root in geographic zones where states have collapsed or are too weak to assert their sovereign authority, and embattled governments believe they can gain an advantage from the influx of foreign fighters. On average, civil wars last longer than international wars, are fought outside conventions regarding the laws of conflict, and are less likely to receive international mediation efforts and reconstruction grants. As a result, they leave a legacy that has been called "development in reverse," because of their long-term economic, health, environmental, and political damage, and their spillover into neighboring countries.[10]

Although the strategic challenge posed by failed states is well-known, policymakers rarely receive adequate warning of their collapse or have the necessary time to effectively marshal domestic and international support for an intervention. Historically, the intelligence community's warning responsibilities have focused on the prevention of surprise military attacks against the U.S. and its allies; the emphasis was on understanding an adversary's organizational routines and military doctrine, and

---

[9] See Ashton B. Carter and William J. Perry, *Preventive Defense* (Washington, D.C.: Brookings, 1999).
[10] See Paul Collier, Lani Elliot, Håvard Hegre, Anke Hoeffler, Marta Reynal-Querol, and Nicholas Sambanis, *Breaking the Conflict Trap* (Washington, D.C.: World Bank and Oxford University Press, 2003), 1-49.

interpreting observable behavior in the context of these variables.[11] Such a focus sought to provide policymakers with advance notice of an impending attack within weeks to days.[12]Post-Cold War, the definition of warning has evolved to include economic, political, and natural disasters.

As the nature of warning has changed, the audience being warned, and the time horizons that policymakers require to make use of warning information have changed, creating new analytic problems.[13] When focused on the Soviet Union, warning analysis concentrated on the factual record of what a target did or did not do, and its emphasis on military warning allowed analysts to reasonably determine what was likely to occur over the near-term based on a generic knowledge of military effectiveness. By comparison, explaining observed political, economic, and social activities is far more complex, in part due to the lack of strong causal models that repeatedly and convincingly link observed behaviors with mid- or long-term outcomes.

Moving away from indications and warning of military operations into broader political and social domains demands the increased use of theories that make sense of available information and observable events and project consequences months or years into the future. However, as warning becomes increasingly dependent upon theory, policymakers can often claim equal or greater expertise than intelligence analysts, meaning that long-term forecasts can be easily dismissed or reinterpreted by those receiving them.[14] Despite the desire for intelligence analysts to offer products that offer greater speculative depth and long-term perspectives, such analyses have been the products that have most often been ignored by consumers.[15] To generate credible warning, PCMT developed an interactive analytic process that allows decisionmakers and their staffs to engage in a dialogue with analysts through the use of multiple models and simulation. Thus, active policymakers, their staffs, and advisors can become part of the analytic process, and any conclusion can be traced back to input data and analytic model, establishing transparency and the ability of users to replicate each other's results to understand how intelligence or policy judgments were reached.

---

[11] Bruce D. Berkowitz, and Allan E. Goodman, *Best Truth* (New Haven, CT: Yale University Press, 2000), 102-104.

[12] See Mark M. Lowenthal, *Intelligence: From Secrets to Policy* (Washington, D.C.: Congressional Quarterly Press, 2000), 87.

[13] See Hayward R. Alker, Ted Robert Gurr, and Kumar Rupesinghe, "The Challenge of Developing Conflict Early Warning Information Systems: A Proposal" in Hayward R. Alker, Ted Robert Gurr, and Kumar Rupesinghe, eds., *Journeys Through Conflict* (Lanham, MD: Rowman & Littlefield Publishers, Inc., 2001), 3-31.

[14] See Sherman Kent, "Estimates and Influence," in Donald Steury, ed., *Sherman Kent and the Board of National Estimates* (Washington, D.C.: Central Intelligence Agency, 1994), 33-42.

[15] See Roger Hilsman, Jr., "Intelligence and Policy-Making in Foreign Affairs," *World Politics*, Vol. 5, No. 1 (October, 1952), 1-45.

National Security and Social Science: Past, Present, and Future

The importance of social science theory to the warning of state failure serves as a reminder of the strained relationship between the social sciences and the national security community. While the physical sciences have played a crucial role in national defense, the relationship between the social sciences and the military has been tense since the events of the 1960s and 70s.

World War II marked a turning point in the relationship between physical scientists and governments with the development of the command economy. The institutionalized relationship between national security and scientific establishments that developed during the war allowed governments to command the creation of new technologies by applying resources to specific scientific fields, issuing contracts and grants and employing other incentives to direct scientific research and education toward specific ends and applications.[16] Although recognized as a source for technological development, World War II also marked the creation of new bodies of knowledge for understanding foreign and domestic public opinion, economics, organizational design, management, training, and communications. During the 1940s, the social sciences played a crucial role in the staffing and training of the military and formed the bedrock of strategic intelligence analysis.[17]

At the outset of the Cold War, the possibility of a cataclysmic nuclear war motivated a new interest in the social and behavioral sciences by strategic planners. Observers noted that patterns of Soviet weapons acquisition did not follow expectations of rational behavior. Whereas classical assessments of states as unified, single actors may have been adequate prior to the Cold War, although not necessarily convincing, misunderstanding an adversary's decisionmaking and behavior in a nuclear stalemate elevated the costs of failure to unacceptable levels. As a result, new efforts to model the international system, the state, and underlying institutions and leaders received considerable attention. Much of the Cold War concept of a balance of terror rested on the use of social science knowledge. Deterrence, the centerpiece of national defense, rested on the understanding of cross-cultural communications, signaling, and deciphering foreign value systems and risk thresholds. Likewise, organizational or operational failures involving the inadvertent handling or release of nuclear weapons, or the accidental signaling of a pre-emptive war were deemed unacceptable and became the focus of

---

[16] See William McNeill, *The Pursuit of Power* (Chicago, IL: University of Chicago Press, 1982); and Merritt Roe Smith and Gregory Clancey, eds., *Major Problems in the History of American Technology* (Boston, MA: Houghton Mifflin, 1998), 427-470.

[17] See Sherman Kent, *Strategic Intelligence for American World Policy* (Princeton, NJ: Princeton University Press, 1949); Roger Hilsman, *Strategic Intelligence and National Decisions* (Glencoe, IL: The Free Press, 1956); Klaus Knorr, *Foreign Intelligence and the Social Sciences* (Princeton, NJ: Center of International Studies, Woodrow Wilson School of Public and International Affairs, 1964); E.S. Quade, ed., *Analysis for Military Decisions* (Santa Monica, CA: RAND, 1964); and David L. Featherman and Maris A. Vinovskis, "Growth and Use of Social and Behavioral Science in the Federal Government since World War II," in David L. Featherman and Maris A. Vinovskis, eds., *Social Science and Policy-Making* (Ann Arbor, MI: University of Michigan Press, 2001), 41-82.

organizational design and training.[18] Indeed, at the beginning of the Cold War, the social science research agenda was so important to national security that one observer noted that if World War II was the physicists' war, World War III (the Cold War) was the social scientists' war.[19]

The collapse of the Soviet Union altered the social science research agenda. Globalization, technological innovation and diffusion, the spread of democracy, cross-civilizational interactions, non-state actors, environmental and demographic change, and the role of moral and ideological factors have largely displaced the Cold War emphasis on state-centric actors and material power.[20] This change has coincided with the development of new analytic tools and theories that relate micro, or individual levels of description to macro, or system level outcomes. These new tools are increasingly important because new challenges, such as terrorism, are not easily reduced to—or understood by—Cold War era analytic methods that assumed stable structures at the national or institutional level.[21] Indeed, while scholars disagree as to the specific dynamics of social systems—the specific rules that relate one level of analysis to another—there exists a strong motivation for studying social dynamics and representing societies in a non-deterministic fashion, where non-structural features such as chance, contingency, and agency play important roles in behavior and outcomes.[22]

---

[18] For examples and assessments of applied social science in the Cold War see Bernard Brodie, *Strategy in the Missile Age* (Princeton, NJ: Princeton University Press, 1965); Andrew W. Marshall, *Problems of Estimating Military Power* (Santa Monica, CA: RAND, 1966); Thomas C. Schelling, *Arms and Influence* (New Haven, CT: Yale University Press, 1966); Graham T. Allison, *Essence of Decision* (Boston, MA: Harper Collins, 1971); Irving L. Janis, *Victims of Groupthink* (Boston, MA: Houghton Mifflin Company, 1972); Morton H. Halperin, with Priscilla Clapp and Arnold Kanter, *Bureaucratic Politics & Foreign Policy* (Washington, D.C.: Brookings, 1974); Alexander L. George and Richard Smoke, *Deterrence in American Foreign Policy: Theory and Practice* (New York, NY: Columbia University Press, 1974); and Bruce G. Blair and John D. Steinbruner, *The Effects of Warning on Strategic Stability* (Washington, D.C.: Brookings, 1991); Marc Trachtenberg, *History & Strategy* (Princeton, NJ: Princeton University Press, 1991), 3-46; Scott D. Sagan, *The Limits of Safety* (Princeton, NJ: Princeton University Press, 1993); Charles Perrow, *Normal Accidents* (Princeton, NJ: Princeton University Press, 1999); and John D. Steinbruner, *The Cybernetic Theory of Decision* (Princeton, NJ: Princeton University Press, 2002).
[19] Seymour J. Deitchman, *The Best Laid Schemes* (Cambridge, MA: MIT Press, 1976), 28.
[20] For a discussion of how the Cold War focused international relations on a narrow set of attributes and capabilities see John Lewis Gaddis, *We Now Know* (New York, NY: Oxford University Press, 1998), 283-284. For an overview of current security threats see Michael E. Brown, ed., *Grave New World* (Washington, D.C.: Georgetown University Press, 2003).
[21] Desmond Saunders-Newton, *Computational Social Science, Operations Research, and Effects-Based Operations: The Challenge of Inferring Effects from Dynamic Socio-Physical Systems*, (McLean, VA: Military Operations Research Society, Workshop on Analyzing Effects-Based Operations, January 29, 2002); and Audrey Kurth Cronin, "Sources of Contemporary Terrorism," in Audrey Kurth Cronin and James M. Ludes, eds., *Attacking Terrorism: Elements of a Grand Strategy* (Washington, D.C.: Georgetown University Press, 2004), 19-45.
[22] See William H. McNeill, *The Global Condition* (Princeton, NJ: Princeton University Press, 1992); W. Brian Arthur, *Increasing Returns and Path Dependence* (Ann Arbor, MI: University of Michigan Press, 1994); Philip E. Tetlock and Aaron Belkin, eds., *Counterfactual Thought Experiments in World Politics* (Princeton, NJ: Princeton University Press, 1996); Lars-Erik Cederman, *Emergent Actors in World Politics* (Princeton, NJ: Princeton University Press, 1997); Claudio Coffi-Revilla, *Politics and Uncertainty* (New York, NY: Cambridge University Press, 1998); G. John Ikenberry, *After Victory* (Princeton, NJ: Princeton University Press, 2001); Randall L. Schweller, "The Problem of International Order Revisited," *International Security*, Vol. 26, No. 1 (Summer, 2001), 161-186; John Lewis Gaddis, *The Landscape of*

Despite their importance to the national security community, the social sciences remain an ignored and neglected area when compared with the physical sciences. One reason for this is their repeated inability to achieve the predictive accuracy found in other fields. A second reason is the cultural divide that emerged between academia and the government during the events of the 1960s and 70s. While PCMT can do little to solve the difficulties of prediction regarding social phenomenon, it is enough to note that the growing field of complexity studies has proven that many problems, and arguably the most interesting ones, in the physical, mathematical, and biological sciences are structurally unpredictable. As a result, the definition of a mature scientific discipline is changing, and prediction as a criterion for measuring scientific progress or value is weakening. Alternatively, PCMT is one small, but important step in rebuilding the relationship between the national security community and social sciences.

The Long Shadow of Project Camelot: Social Science and Defense Research

Project Camelot, a U.S. Army research project focused on insurgency and political revolution in the 1960s, reveals the difficulties of applying social science research to security policy and serves as one of the defining moments in the relations between social scientists and the military. Camelot was a large-scale research program focused on identifying the sources of social change and the development and success of insurgencies. Camelot was intended to produce peer-reviewed, published, academic papers and data and a theory of social dynamics, and to investigate the potential of a computational model for simulating social and political change. This work was to be performed by academic researchers under contract to the U.S. Army conducting archival research, and some field studies in foreign countries. Their results were intended to be used by the Army for developing counterinsurgency doctrine to combat Soviet sponsored "wars of national liberation." However, a series of embarrassing, though false, accusations were made about the activities of the program, and it was ultimately cancelled after less than one year. While the motivations and views of Camelot's critics varied, the most important and enduring criticisms dealt with the proper role of the military in the development of social knowledge, the institutional relationship between the military and the academic community, and the practical and moral complexities with applied social science and social engineering.[23]

Camelot cast a long shadow. The Department of Defense was prohibited by Congress from performing research that did not have a clear military application. While the majority of DOD research programs were easily repackaged and continued, nearly all social science research ceased, except in the cases of deterrence, training, and organizational studies. The vision of the social sciences providing powerful, proven tools for states to harness in their foreign policy never materialized, despite many efforts to

---

History (New York, NY: Oxford University Press, 2002); and Duncan J. Watts, *Six Degrees* (New York, NY: W.W. Norton & Company, 2003).

[23] For a history on the project and the issues surrounding it see Irving Louis Horowitz, *The Rise and Fall of Project Camelot* (Cambridge, MA: MIT Press, 1974); and Seymour J. Deitchman, *The Best Laid Schemes* (Cambridge, MA: MIT Press, 1976).

apply leading theories and align research agendas with desired applications.[24] Moreover, many of the underlying controversies about the causes of political or social phenomena remain unresolved. For example, does terrorism result from a deficit of liberal, democratic institutions, or is it a response to the emergence of those institutions and the accompanying erosion of traditional, social, economic, and political systems? Is state sponsorship a prerequisite for the successful planning and execution of terrorist acts, or can effective groups form and merge without access to national resources and covert or overt government sponsorship?[25]

PCMT and the Social Sciences: Rebuilding Bridges

After the embarrassment of Project Camelot, and the subsequent events of Vietnam and Watergate, the public's trust and faith in the military and government declined, and the gap between academic researchers and the government widened. PCMT seeks to rebuild the bridge between the government and social sciences. Its objectives are similar to those of Camelot, but differ in two important ways.

First and foremost, PCMT is designed to work with data that is publicly available, or with datasets that its users wish to use with the tools. Unlike Camelot, PCMT has no organic means for performing field research or tasking the collection of specific information. PCMT can search through publicly available documents, and can accept any information that its users wish to use with the tools, but it has no programmatic authority to turn researchers into government agents. Therefore, a programmatic wall exists between the collection of data and the use of information that preserves the independence and integrity of each activity.

Second, whereas Camelot sought to settle disputes within the social sciences, effectively producing a government-endorsed theory of social behavior, PCMT has no such agenda. Instead, PCMT is focused on providing users with a diverse collection of social science models and theories. By not attempting to create a new theory, or even become immersed in theoretical arguments within the social sciences, PCMT is able to employ models that have been developed and used by researchers within the norms of their communities, using the peer review process as a means for model validation. Moreover, PCMT emphasis on disciplinary, theoretical, and methodological breadth enables its analytic suite to draw from political science, economics, sociology, anthropology, and other disciplines that employ both quantitative and qualitative models to provide users diverse perspectives. Thus, PCMT emphasizes model use and simulation—rather than the accuracy of a specific model or preferred theory—to show

---

[24] See Charles W. Bray, "Toward a Technology of Human Behavior for Defense Use," *The American Psychologist*, Vol. 17 (1962), 527-541; and Michael E. Latham, *Modernization as Ideology* (Chapel Hill, NC: The University of North Carolina Press, 2000).

[25] Such questions were supposed to be addressed by Project Camelot, but remain unanswered to this very day. For examples of how this dialogue has continued see Martha Crenshaw, ed., *Terrorism, Legitimacy, and Power* (Middletown, CT: Wesleyan University Press, 1983); David C. Rappoport and Leonard Weinberg, eds., *The Democratic Experience and Political Violence* (Portland, OR: Frank Cass, 2001); Audrey Kurth Cronin, "Behind the Curve: Globalization and International Terrorism," *International Security*, Vol. 27, No. 3, 30-58; Michael Mousseau, "Market Civilization and Its Clash with Terror," *International Security*, Vol. 27, No. 3 (Winter 2002/2003), 5-29; and John Lewis Gaddis, *Surprise, Security, and the American Experience* (Cambridge, MA: Harvard University Press, 2004).

users how different perspectives on a problem may agree or disagree regarding the international system's future trajectory, and establish a landscape of potential futures from which policy options should be evaluated.

By pursuing the development of a single model, Camelot sought to resolve long-standing theoretical disputes within the academy, effectively making the military and intelligence community arbiters in methodological and theoretical disputes. By comparison, PCMT seeks to keep the government out of such debates. Instead, its agenda fundamentally alters the relationship between policy and research by encouraging and supporting the development and employment of multiple, competing hypotheses. Rather than search for a single, perfect model, which burdens users with the requirements of multiple forms of model validation, the PCMT decision-support methodology emphasizes responsible model usage and inferencing, i.e., reasoning from models.[26] As a result, the practice of searching for policies or strategies that are optimized around the predictions of a single model are eschewed in favor of polices that are robust and perform well across a collection of models based on satisficing across potential futures.

The PCMT Program hopes to provide a forum that encourages the use of, and interest in social science research and models—to lay a foundation that allows the government to be a responsible user of social science research and models, without exerting pressure on the development of social theories, or undermining the objectivity of academic researchers. Turning the government into a consumer of social models and theories should give social scientists increased incentives to study problems and issues that are relevant to policymakers.

---

[26] For a discussion of different forms of model validation in the social sciences see Charles S. Taber, and Richard J. Timpone, *Computational Modeling* (Thousand Oaks, CA: Sage Publications, 1996), 71-79.

The PCMT Architecture

From a programmatic perspective, PCMT is divided into three subsystems: data collection, analytic modeling and simulation, and interagency collaboration infrastructure. Each subsystem is capable of working independently, but their effectiveness greatly increases when employed as a whole.

A New World of Information: Automated Data Collection and Organization

The advent of digital communications, the digital production and manipulation of information, and the networking of information systems lie behind an information explosion. Discrete sources of information are merging. People and systems that worked in isolation are now locally and globally connected. A massive, unstructured corpus exists in the form of the Internet, which is constantly expanding;

- Between 1999 and 2002 the total amount of information produced per year grew by 30 percent;
- Five exabytes of data are produced per year, the equivalent of 500,000 Libraries of Congress;
- 92 percent of all information is stored on some form of magnetic media;
- The World Wide Web (WWW) contains 170 terabytes, as much information as 17 Libraries of Congress;
- The source for the largest volume of information flows is the telephone network; the Internet is second;
- Text, audio, and video are increasingly originating and being distributed in digital form.[27]

PCMT confronts this universe with an integrated approach that is designed to collect relevant information without drowning analysts in overwhelming amounts of data. PCMT methods of data collection and document coding provide significant savings in terms of time and money. As a result, analysts can spend more time thinking about complex problems and performing analysis, rather than managing information, updating datasets, and arranging information for use in mental, computational, or mathematical models.

Traditionally, analysts extract information about a given problem by sampling from the available universe of data — they examine a small portion of the information with the expectation that the sampled portion accurately represents the character of all available data. However, as the quantity of information about the world has increased three changes to the nature of information and analysis have occurred.

First, the universe of information has expanded. The Internet has enabled researchers to access foreign newspapers, blogs, television and radio broadcasts, NGO and IGO reports, corporate and investor development plans, etc. Indeed, analysts confront

---

[27] Peter Lyman, Hal R. Varian, James Dunn, Aleksey Strygin, and Kirsten Swearingen, *How Much Information? 2003* (Berkley, CA, University of California at Berkley, 2003) available at http://www.sims.berkeley.edu/research/projects/how-much-info-2003/.

massive quantities of information whose quality and formats are vastly broader than they were only a decade ago. If analysts confine their sampling to familiar sources, they risk neglecting an ever-expanding array of sources that change the overall character of the dataset.

The second change brought on by the explosion of information has been the move toward the digital production and distribution of information. Processes that required entire studios to produce or publish years ago can be embedded into single microchips or software. Moreover, as the availability of microcomputers has expanded, individuals and classes of people who were formerly excluded from social, political, economic, scientific, or military discourses can now be active participants. As the costs of information production and communication decline and the number of participants on the Internet increase, society's sensory organs grow denser, and the population of politically active and aware people increases.[28]

The third result of the information explosion is the development of new methods for categorizing and searching data. New search tools provide the means for understanding the character of, and trends within, large datasets. While sampling may not be adequate, given the changing character of data, filtering methods—the employment of pattern matching and templates to examine all available data—are becoming increasingly mature and important for understanding how individual pieces of data relate to, or represent the character of the larger universe of information.

PCMT employs several data collection techniques—ranging from automated agents searching for specific pieces of information to accepting unstructured contributions from a user community—to gather and categorize a broad range of data for use in its analytic suite. Not only is the information in the PCMT database dynamic, but its structure is adaptable and can be customized by any user according to the specific ontology they desire. The shift toward filtering can alert analysts to important trends within the dataset, indicate when small data samples may not reflect the character of the larger universe, or identify individual documents or sources that are outliers in terms of content, tone, timeliness, or other features that may be of interest.

PCMT automated data collection and categorization tools allow organizations to conduct analysis in ways that are currently beyond the scope of their resources. Indeed, a small demonstration of PCMT capabilities reveals the power of automated data collection and coding systems, and the subsequent consequences for organizational budgets, staffs, and production cycles. An automated data collection activity focused on news articles during the period between August 2003 and January 2004 averaged more than 40,000 English language articles per month on Afghanistan, 37,000 for Kazakhstan, and 27,000 for Pakistan. Indeed, when dealing with the country with the smallest number of articles, Turkmenistan, automated data collection tools were still able to collect more than 12,000

---

[28] For more on the notion of increasing sensor density as a function of declining cost see Martin C. Libicki, *The Mesh and the Net* (Washington, D.C.: National Defense University Press, 1994), 19-50. For a view of the increasing density of political and social interactions see William H. McNeill, "The Changing Shape of World History," *History and Theory*, Vol. 34, No. 2 (May 1995), 8-26; and Claudio Cioffi-Revilla, *The Big Collapse: A Brief Cosmology of Globalization* (Fairfax, VA: Center for Social Complexity, George Mason University, July 2004).

articles a month. In another case, PCMT "webscraping" tools collected over 60,000 documents on Central Asia in a single day.[29]

Depending on the complexity of the model and the document, e.g., newswires, magazines, journal articles, or books, experienced humans can code a document for use in a model in as little as twelve minutes.[30] Based on this rate, a full-time staff of 180 people would be required to code all available news articles on Central Asia on an annual basis. At an estimated cost of $100,000 per man-year of labor, the financial cost of covering Central Asia would be $18,000,000. Few government organizations can afford such costs in time, labor, or budget, so such methods of analysis are rarely employed, and then only on small datasets over short temporal durations.

PCMT automated data collection and document-coding tools enable organizations to conduct analysis that was previously too costly to perform. Experiments conducted by JFCOM have demonstrated that adopting a machine assisted coding system reduced the time required to code a document by 80% and needed staff from 180 people to 36, at a projected personnel cost of just over $3,500,000, rather than $18,000,000. PCMT is making further improvements in its coding technology. When the full machine coding system is implemented, manpower and organizational costs will be reduced by another 80%, meaning that all collected documents on Central Asia can be coded with a staff of seven people at an estimated cost of under $1,000,000 dollars.

PCMT automated data collection and document coding tools enable analysts to employ methods and models that would otherwise be too time consuming and costly. Furthermore, they allow social science models to work from disaggregated data collected in near-real time—daily or weekly intervals—in addition to more traditional sources published monthly, quarterly, and annually.[31] The ability to work with new data sources, better understand the character of available data, and rapidly process incoming data for use in multiple modeling schemes provides the foundation from which PCMT supports analysts and decision-makers.

Analytic Modeling and Simulation

The PCMT analytic suite provides users with an understanding of the origins and drivers of violent conflict to help them preventing its outbreak. The analytic suite combines multiple models of social vulnerabilities, trends, and elite dynamics to craft and implement robust and adaptive policies. Through the use of competing models and multiple data sets, landscapes of scenarios or potential futures are generated. From these landscapes, decisionmakers can simultaneously address multiple futures by identifying and implementing strategies that perform well across the entire landscape, building in hooks, or decision-points, for refinement and adjustment as time passes and the landscape changes.

---

[29] Figures provided by the Institute for Physical Sciences and DARPA.

[30] Figure provided by the Fund for Peace based on experience with the Conflict Assessment System Tool.

[31] Gary King and Will Lowe, "An Automated Information Extraction Tool for International Conflict data with Performance as Good as Human Coders: A Rare Events Evaluation Design," *International Organization*, Vol. 57 (Summer 2003), 617.

Social Vulnerability Models

The causes of war are often debated. Multiple theories and perspectives exist on the sources of conflict, and disagreements over the conditions and structures that give rise to violence are likely to persist.[32] Although immersing decisionmakers in theoretical debates regarding the causes and nature of war may prove enlightening, it would be counter-productive to give them the burden of resolving these disagreements. Alternatively, presenting conclusions or judgments based on the assumptions or applications of a single theory masks important disagreements within communities of experts, creating the illusion of certainty where none exists. PCMT addresses the problem of competing theories by providing a suite of analytic models that covers a gamut of disciplinary and methodological perspectives on violent conflict. From this suite, policy-makers receive diverse and independent assessments of a situation, enabling them to rapidly examine a landscape of potential futures without having to become experts in individual models or theories.

At this time, the PCMT analytic suite employs two social vulnerability models that assess the likelihood a state may descend into civil war. The first model is the Conflict Assessment System Tool (CAST) developed by the Fund for Peace.[33] CAST has been used by government agencies for conducting regional assessments and planning for over a decade. The second model is based on the work of Paul Collier and his colleagues at the World Bank Group, and has been used for estimating the risks associated with country loans and determining the consequences of development on conflict potential by the World Bank.[34] Each of these models are focused on estimating the likelihood of civil war or state failure, yet their methodological approaches and underlying theory are quite different.

Conflict Assessment System Tool. The CAST model forecasts a state's risk of failure. The general theory the model instantiates is that conflict is a process that begins with the decline of a state's governing capacity.[35] The CAST model tracks twelve indicators that monitor the internal health of a state. These indicators are:

- Demographic Pressures
- Internal Displacement of People
- Legacy of Vengeance Seeking Groups
- Chronic Human Flight
- Unfair Economic Development
- Severe Economic Decline
- Criminalization or Deligitimization of the State

---

[32] For a testament to the enduring uncertainties regarding the causes of war compare Quincy Wright's study of war from the 1920s with Greg Cashman's conducted in the 1990s. Despite changes in methodology, the occurrence of major historical conflicts, and the growth of the field during the Cold War, little resolution to the causes of conflict has occurred. See Quincy Wright, *A Study of War* (Chicago, IL: University of Chicago Press, 1982); and Greg Cashman, *What Causes War?* (Lanham, MD: Lexington Books, 1993).

[33] The Fund for Peace is a non-profit, think tank focused on conflict prevention based in Washington, D.C. that focuses on the problems of weak, failing, and failed states. Detailed information about the Fund for Peace is available at http://www fundforpeace.org.

[34] Within the PCMT lexicon this World Bank model is referred to as the Collier model or Collier for short.

[35] See Fund for Peace, *CPR Model: II-B. Theoretical Assumptions* (Washington, D.C.: Fund for Peace), available at http://www fundforpeace.org/programs/cpr/assumes.php.

- Deterioration of Public Services
- Violation of the Rule of Law and Human Rights
- Security Apparatus Independent of the State
- Rise of Factionalized Elites
- Intervention of External Political Actors

Each indicator comprises several measures based on data items representing quantitative measurements, e.g., birth and death rates, and qualitative judgments, e.g., whether the population considers the judicial system fair and responsive. Qualitative judgments and other normative assessments are evaluated according to existing international law and norms.[36] A state's risk of failure is assessed based on the measurements and subsequent indicator scores. In total, CAST employs more than 4,000 data items, many of which affect multiple indicators. Each of the twelve indicators is evaluated by averaging the scores assigned to their associated data items on a 0-10 scale.[37] A state's risk of failure is then calculated based on the sum of each indicator's value. The higher the total score, the more likely the occurrence of state failure.

In the CAST model, as governing capacity declines, the state's ability to deal with demographic, economic, security, environmental, and public health problems diminishes. As a result, competing groups form, each seeking to provide the services that the government cannot. Such groups may be organized around ethnic, regional, religious, or commercial lines. While the emergence of these groups may signify a strong civil society that can provide services at a grass roots level, if there is a high degree of competition between them and questionable legitimacy, conditions for factional conflict increase. Additionally, competition over natural resources, the influx of cash, arms, and ideology from diasporas, and the provocations of foreign governments may encourage violent escalation. Thus, CAST views civil conflict as a process with distinct decision points or branches, indicating future conflict trajectories and opportunities for intervention to reestablish order and bring warring parties to peaceful terms.[38]

Inputs for the CAST model consist of all available documents within the PCMT database. They include quantitative and qualitative, and structured and unstructured data that has been collected via automated agents or submitted by members of the user community. This data is analyzed using machine-assisted and machine-coding methods discussed above, and then passed directly into the modeling software. Outputs include risk scores based on the total assessment of all indicators, as well as indicator-specific scores and assessments of the governing capacity of basic legislative, executive, judicial, and military institutions. Trend analysis is available, as is the ability to drill-down into documents, data-items, and assigned scores that determine model results.

The Collier Model. The Collier Model is a quantitative model based on the work of Paul Collier and his colleagues at the World Bank Group that has been employed to

---

[36] Discussion with Pauline Baker from Fund for Peace.

[37] The original CAST model employed a scale of –5 through +5. This numeric scale has been changed to a 0-10 scale for the purposes of creating a computational model.

[38] For a detailed explanation of the CAST model and theory behind its construction see Pauline H. Baker and John A. Ausink, "State Failure and Ethnic Violence: Toward a Predictive Model," *Parameters* (Spring 1996), 19-31; and the *Conflict Prevention and Recovery Program* at the Fund for Peace website at http://www.fundforpeace.org/programs/cpr/cpr.php.

understand and articulate the relationship between development and conflict.[39] It is actually a collection of statistically indistinguishable models that emphasize different drivers of civil conflict. In one case, the model assumes that civil wars result from political or social grievances within the population, and that conflicts are political in character and motivation. In another case, the model assumes that civil wars are driven by greed—profit and rent seeking activities by elites within the state—and that the drivers of conflict are economic and tend to be clustered around resources or other economic assets. Other cases are based on the inclusion or exclusion of particular variables such as the duration of time between conflicts within each state.[40]

Both the grievance and greed models indicate that civil war is development in reverse and track important development indicators, such as mortality rates, infrastructure, and foreign investment before and after conflict. A crucial finding of the Collier model is that civil wars are self-reinforcing and likely to recur, absent intervention in their economic, political, and social legacies.

The Collier model is based on a logistic regression of data on 161 countries over the period between 1960 and 1999. The regression calculates the probability of civil war in a given country based on the categorization of variables into three groups. The first group is the most recent set of country indicators — GDP, commodity exports, population characteristics, etc. The second group of variables is from the preceding time period, and is used to calculate change over time, e.g., the growth or decline of per capita income. The final group of variables refers to structural factors that do not change over time, or change slowly over generations. The data employed by the model include variables such as:

- Diaspora populations
- Ethnic dominance
- GDP per capita
- Geographic dispersion of population
- Peace duration
- Population
- Primary commodity exports/GDP
- Social, ethno-linguistic, and religious fractionalization
- War start[41]

PCMT has employed the Collier Model due to its substantive focus on civil war and its acceptance within the international community and actual application within the World Bank.

---

[39] See http://econ.worldbank.org/programs/conflict/.

[40] Paul Collier and Anke Hoeffler, *Greed and Grievance in Civil War* (World Bank Working Paper, October 21, 2003) available at http://www.wds.worldbank.org/servlet/WDS_IBank_Servlet?pcont=details&eid=000265513_20040310152555.

[41] Paul Collier, Lani Elliot, Håvard Hegre, Anke Hoeffler, Marta Reynal-Querol, and Nicholas Sambanis, *Breaking the Conflict Trap* (Washington, D.C.: World Bank and Oxford University Press, 2003), 189-196.

Elite Dynamics

PCMT social vulnerability models provide insights into a society's structural condition and trends regarding political stability. To complement the use of information provided by these models is the use of elite dynamics and the examination of network structures and relationships between leaders, organizations, and issues. The inclusion of elite dynamics into the PCMT analytic suite serves three purposes. First, it enables an understanding of what policy options are viable based on existing relationships within the region. Second, elite dynamics serve as leading indicators of important events or emerging issues within a society before they manifest in other forms. Finally, elite structures can be employed as indications of policy success or failure, serving as a measure of effectiveness for national strategy and international action.

PCMT use of elite dynamics rests atop a large literature and practice of thinking about states and leaderships as a system of interconnected parts.[42] Within the context of PCMT, elite dynamics is understood as a means for observing how societies extract and mobilize mass resources and how hierarchies of decisionmakers emerge in formal and informal political organization.

The examination of elite networks and information flows, whether between individuals or organizations, issues, and the unfolding of events has been used to depict and understand political relationships and open up the traditional black boxes of party, race, gender, class, nation, and other aggregate levels of analysis. While aggregates provide a useful mechanism for arranging a system into its constituent elements, large collections of people and organizations assume the uniformity of their members and do not account for the relational properties and the consequences of micro-level interactions and processes within groups.[43] The analysis of elites helps explain the methods by which political, social, economic, and military resources are extracted, concentrated, and deployed within a society, and how the mass mobilization of people, for or against a particular policy, occurs.[44] In addition, network analysis methods are critical for developing alternative models of the international system that go beyond the examination of regional blocks and the viewing of nations as unitary actors. This is important for understanding the construction and dynamics of a post-Westphalian world, where groups may form and dissolve dynamically depending on opportunities and threats, and how

---

[42] For examples see Klaus Knorr and Sidney Verba, eds., *The International System* (Princeton, NJ: Princeton University Press, 1961); Thomas Rona, *Our Changing Geopolitical Premises* (New Brunswick, NJ: Transaction Books, 1982); Bernard I. Finel and Kristin M. Lord, eds., *Power and Conflict in the Age of Transparency* (New York, NY: Palgrave, 2000); Barry Buzan and Richard Little, *International System in World History* (New York, NY: Oxford University Press, 2000); James N. Rosenau and J.P. Singh, *Information Technologies and Global Politics* (Albany, NY: State University of New York Press, 2002); J.R. McNeill and William H. McNeill, *The Human Web* (New York, NY: W.W. Norton & Company, 2003); and Shanthi Kalathil and Taylor C. Boas, *Open Networks Closed Regimes* (Washington, D.C.: Carnegie Endowment for International Peace, 2003).

[43] See David Knoke, *Political Networks* (New York, NY: Cambridge University Press, 1994), 1-27.

[44] See David Knoke, *Political Networks* (New York, NY: Cambridge University Press, 1994), 153-155; and Nematollah Nejoumi, *The Rise of the Taliban in Afghanistan: Mass Mobilization, Civil War, and the Future of the Region* (New York, NY: Palgrave, 2002).

collections of actors arrive at consensus about abstract issues or concerns for which no empirical or objective criterion for evaluation exist.[45]

To appreciate the importance and application of elite dynamics within the PCMT analytic suite, it is useful to consider two analogous applications of elite analysis: marketing and development. One of the first uses of network analysis was in the 1950s, when pharmaceutical companies grew interested in understanding how new prescription drugs entered and propagated within markets. Researchers traced the adoption of a new antibiotic by studying which doctors were the first to prescribe it to their patients, which of their peers were the first to emulate them, and who were the last to prescribe the antibiotic. Since that initial study five decades ago, the examination of network structures, and the actions of the individuals within them, has been used to determine whether an innovation or idea will go critical and spread across the entire network, remain isolated within small isolated cliques, or die out.[46]

As in marketing, network analysis has assisted the area of international development. Elite networks have been used to explain the success of microloans, microcredit, and microfinance programs, where small groups or individuals at very local levels receive loans or aid. By working through local actors who are well placed in social or community networks, community leaders become partners in the development process.[47]

In both the marketing of new products and the distribution of small loans and grants, society's structure and the relationships between authority figures within it reveal information about potential futures and the emergence, acceptance, rejection, and long-term viability of various practices, beliefs, and policies. PCMT analysis of elite dynamics is designed to provide users with the policy equivalent of network-based marketing or microloan tools, assisting policymakers in developing messages and undertaking actions that are compatible with existing and emergent social and political structures.

Understanding how networks operate at national and local levels enables the crafting of policies that are appropriately nuanced and tailored for specific regions, countries, or indigenous partners and allies. Coupling networks of elites with the information derived from social vulnerability models provides PCMT users with a systematic method for creating partnerships within the region by working with individuals who have a vested interest in increasing and maintaining political and social stability, and the ability to direct and perform actions at local levels to prevent the outbreak of violence.

Decision-support

PCMT employment of social science models for decision-support presents users with a series of advantages and challenges compared with more traditional operations

---

[45] See Patrick Doreian, "Models of Network Effects on Social Actors," in Linton C. Freeman, Douglas R. White, and A. Kimball Romney, eds., *Research Methods in Social Network Analysis* (New Brunswick, NJ: Transaction Publishers, 1992), 298; David Knoke, *Political Networks* (New York, NY: Cambridge University Press, 1994), 202; and Robert Axelrod, *The Complexity of Cooperation* (Princeton, NJ: Princeton University Press, 1997), 124-147.

[46] See Albert-Laszlo Barabasi, *Linked* (Cambridge, MA: Perseus, 2002), 123-142.

[47] Jonathan Mordach and Manohar Sharma, *Strengthening Public Safety Nets from the Bottom Up* (Washington, D.C.: The World Bank, September 2002).

research tools used in tactical or operational planning. To understand these advantages and challenges, it is important to consider several features of policy analysis and the models that support its development and implementation–the validation of models, inferencing and drawing meaning from their outputs, and the subtle distinctions between prediction, projection, and exploration.

Model Validation. Models are built to help users understand some facet of a problem or phenomenon. By design, models accentuate some features of the problem or phenomenon they represent while ignoring others. Thus, no model is a perfect representation of reality; it is a tool for helping users understand the world, not replications of it. Therefore, the intended use of a model is intimately and inseparably linked with its design and the selection of what features to bring to the forefront and what to conceal. For example, a life-size map would allow for the most realistic and detailed representation of the drive from California to New York, but it would be a useless to a driver planning the trip.[48] Because all models are representations of reality, it is necessary to consider under what conditions their features allow for a valid, credible representation of it. At some level, all models fail to represent the real world in one way or another; they are not isomorphic with reality and are what one analyst called "bad models."[49]

Traditionally, the analytic communities have asserted their need for validated models, i.e., models whose outputs predict behaviors that are consistent with empirical observations. However, the emphasis on prediction has biased the community toward depicting problems in ways that are easily represented and bounded, limiting the scope of inquiry to phenomena that are fundamentally predictable and making caricatures of more complex phenomena that cannot be easily represented mathematically.[50] The over-simplification of problems to enable their modeling has produced a counterproductive analytic effect resulting in models that are predictive and validated within frameworks that bear little resemblance to the real world and provide insights that are irrelevant, too vague, or simply unhelpful to real decision-makers.[51] As a result of using prediction as the sole criterion for validation, users have relegated models to the support of operations planning, while mostly ignoring the more complicated, interesting, and important problems of strategy and policy.[52]

When dealing with complex, social phenomena, model validation requires a more complete notion of the linkages between theory, model, and reality. The relationship between theory and model is one of verification and internal validation. Verification ensures that a model is internally consistent in that the workings of the model reflect the features that the analyst intended. Stated differently, a verified model is one that the developer intended to build and that accurately and faithfully represents the theory upon

---

[48] Edith Stokey and Richard Zeckhauser, *A Primer for Policy Analysis* (New York, NY: W.W. Norton & Company, 1978), 8-21; and John Lewis Gaddis, *The Landscape of History* (New York, NY: Oxford University Press, 2002), 31-34, 45-46.

[49] See James S. Hodges, *Six (Or So) Things You Can Do with a Bad Model* (Santa Monica, CA: RAND, 1991), 1-2.

[50] Charles S. Taber and Richard J. Timpone, *Computational Modeling* (Thousand Oaks, CA: Sage Publications, 1996), 2-11.

[51] John D. Steinbruner, *The Cybernetic Theory of Decision* (Princeton, NJ: Princeton University Press, 2002), 327-331.

[52] Herbert A. Simon, *Administrative Behavior* (New York, NY: The Free Press, 1997), 21-23; Herbert A. Simon, *The Sciences of the Artificial* (Cambridge, MA: MIT Press, 2001), 28.

which it rests.[53] Internal validity goes further to assert that the theory itself is credible and that its assumptions or axioms are not contradictory, such as modeling humans operating under the conditions of bounded rationality while endowing them with superhuman information processing capabilities.[54]

The more frequent use of validation refers to the notion of external validation, the process by which a model's outputs are compared with the phenomena under consideration within the constraints of the model's intended application. External validation relates the model to the real world. In cases where models examine phenomena in which outputs can be tested against empirical evidence, such as in simulating the characteristics of a weapon's design and immediate physical effects, the validation process provides users with the ability to use the model for predictive purposes.[55] In other cases, models may be used for the management of information, in automated management systems, in the development of *a fortiori* arguments, as an aid to theory building and hypothesizing, as an aid in selling ideas for which the model is a conceptual illustration, as a training aid to induce particular behaviors, and, most importantly from the perspective of PCMT, as a decision-aid in operational settings.[56]

Because PCMT employs models to help users consider potential futures, there is no empirical record of the future with which to compare simulation results. While retrodiction is commonly used as an alternative to prediction in such cases, the practice has inherent methodological weaknesses, such as assuming the equivalence of past and future cases, the advantaged treatment given to historical data, and an inability to adequately deal with stochastic outcomes.

When assuming equivalence between cases, the modeler asserts that the future will be like the past in all the dimensions that the model considers. However, a model specifically tailored to given cultural, technological, political, economic, military, or other conditions may perform quite well within one context, but poorly in others. Because no model is isomorphic with reality, the generalizability of results is limited by degree. Moreover, the rise of new actors and conditions is part of the historical process, meaning that often no historical analogue or comparable case exists. A model tailored to the analysis of states such as those in modern Central Asia, that have only existed since the 1990s, may rely on social, cultural, governmental, military, economic, and other factors that were nonexistent only or decade or two before, making comparative assessment impractical from the perspective of determining the model's validity.

In the case of advantaged data, historical databases naturally clean and prune data, and new knowledge fundamentally alters knowledge about cause and effect. Knowing that the allies cracked the Enigma machine during World War II, that the Soviets had deep penetrations into U.S. intelligence, such as Aldridge Ames, or that the Soviet

---

[53] E.S. Quade, *Analysis for Public Decisions* (New York, NY: North-Holland, 1984), 152; and James S. Hodges, *Six (Or So) Things You Can Do with a Bad Model* (Santa Monica, CA: RAND, 1991), 2.

[54] Charles S. Taber and Richard J. Timpone, *Computational Modeling* (Thousand Oaks, CA: Sage Publications, 1996), 77-79.

[55] Peter D. Zimmerman and David W. Dorn, "Computer Simulation and the Comprehensive Test Ban Treaty," *Defense Horizons*, No. 17 (August 2002), 1-5.

[56] For a detailed accounting of these purposes see James M. Hodges, *Six (or So) Things You Can Do with a Bad Model* (Santa Monica, CA: RAND, 1991), 3-8; and James S. Hodges and James A. Dewar, *Is It You or Your Model Talking? A Framework for Model Validation* (Santa Monica, CA: RAND, 1992), 19-31.

economy was far smaller than realized throughout the Cold War forever changes one's view of events and any effort to model them. The conditions under which past events are modeled and compared with current ones are not constant, further complicating validation by retrodiction.

Finally, because history is stochastic, and small and random events can exert significant influence, determining whether a model is a good fit is not a straightforward endeavor. Consider two models that predict a war between two states with some frequency but assume vastly different causation. Meanwhile, a third model, using a causal explanation different from the other two, frequently predicts peace, but occasionally predicts war. Which model, if any, is valid if the prediction of war is historically correct? Because events can only unfold once, it is not necessarily the case that the model that predicts war with greater frequency is a better representation of the real causal process than the model that predicts war occasionally. Despite this problem, popular validation efforts regularly use correct historical prediction as a means of validation, thereby biasing the representation of history as structurally determined and downplaying the effects of agency, chance, and contingency.[57]

Given the complexities of validation based on retrodiction, PCMT does not consider it an essential feature of the modeling suite, although some of its models, e.g., CAST, have used retrodiction for the purposes of validation. Instead, PCMT has left it to the model designers and academic community to determine the validity of specific models through standard best-practice and peer-review processes that adhere to the norms and constraints of the social science community. While this does not guarantee the predictive accuracy of the models themselves, it ensures that they are the state-of-the-art within their respective disciplines and communities. Although these models do not meet the more conservative validation criteria of prediction, these models retain their validity based on the linkage between their internal design, inherent capability, and intended use in the decision-making process.

Inferencing and Credibility. The actual use of models for analytic purposes, whether predictive, exploratory, theoretical, or otherwise is an act of inferencing. It is important to remember that information derived from a model's outputs is a statement about the model and its internal design and processes, not reality. The model itself consists of a series of propositions about how to manipulate the information within it, whether in the form of data, objects, processes, or a combination of the three. Thus, the act of inferencing is the process through which model users given meaning and value to the model's outputs.

Credibility is related to validity in that it ensures that models are used in a fashion consistent with their inherent capability, and that inferencing from outputs does not assert the model is manipulating variables or performing calculations beyond what its internal structure allows. As the nature of the phenomenon under investigation increases in complexity, the credibility of model outputs change. Models that depict the ballistics of a bullet are far different than models that depict the consequences of gun ownership on

---

[57] See Lars Erik-Cederman, "Rerunning History: Counterfactual Simulation in World History," in Philip E. Tetlock and Aaron Belkin, eds., *Counterfactual Thought Experiments in World Politics* (Princeton, NJ: Princeton University Press, 1996), 247-267; and John Lewis Gaddis, *The Landscape of History* (New York, NY: Oxford University Press, 2002).

society and political processes. The number of variables, the relationships between them, and the character of the individual components differ, and therefore the predictability of the underlying phenomena differs as well. Given that the models included in the PCMT analytic suite address social, political, economic, and environmental conditions careful inferencing, and the credible application of the tools at hand are essential features of deriving valuable warning and decision-support to policy-makers.

  Prediction, Exploration, and Anticipation. A final feature of model use is the question of prediction. As previously mentioned, the traditional model validation process consists of comparing model outputs with empirical evidence. However, in many cases model outputs, or even the model itself, cannot be compared against reality in a factual sense, yet the models retain intellectual value and policy relevance, e.g., the model of a perfect market in economics. Perfect markets do not exist in reality, yet as a model of economic behavior they are useful for providing insights into the consequences of different policy options.

  Based on different types of outputs and degrees of model responsiveness, consideration for whether a model is predictive in a mathematical sense, explores the probabilistic likelihood of a specific outcome, or makes an untestable statement about the future affects how the model should be employed and in what parts of a decision-making process it can be used. In a mathematical sense, prediction refers to a point estimate, a singularity in space or time. In another sense, predictive models have a known degree of predictive success, i.e., the accuracy of a model's predictions is known to be correct with a given frequency.[58] Unfortunately mathematical predictions or point estimates are simply inappropriate and fail to address the complex reality of non-deterministic phenomena that PCMT assists decision-makers in understanding. Indeed, efforts to define the problems of civil war or state failure into simple, linear, predictable models result in obfuscating the most interesting and important features of the problems on which policy is focused.

  The social vulnerability models employed by PCMT are what has been referred to as weakly predictive and fare better when used for exploratory analysis or anticipation.[59] In these cases, model outputs are not viewed as singularities, but rather suggestive outcomes. Whereas a ballistics model may provide an output related to a bullet's trajectory, an outcome that can be tested in an experimental setting such as a firing range, models of more complex processes cannot be evaluated in such a fashion. In cases where models deal with phenomena that occur once, or at least infrequently — such as historical outcomes — the dogmatic comparison of model outputs with empirical evidence as a method of validation undermines the ability of analysts to understand the effects of chance and contingency, and identifying critical paths within the unfolding of complex phenomena.[60]

---

[58] See James A. Dewer, Steven C. Bankes, James S. Hodges, Thomas Lucas, Desmond K. Saunders-Newton, and Patrick Vye, *Credible Uses of the Distributed Interactive Simulation (DIS) System* (Santa Monica, CA: RAND, 1996), 23-24.

[59] James A. Dewer, Steven C. Bankes, James S. Hodges, Thomas Lucas, Desmond K. Saunders-Newton, and Patrick Vye, *Credible Uses of the Distributed Interactive Simulation (DIS) System* (Santa Monica, CA: RAND, 1996), xv-xvi.

[60] See John Lewis Gaddis, *The Landscape of History* (New York, NY: Oxford University Press, 2002), 91-109.

The fact that models of conflict, such as those employed in the PCMT analytic suite, cannot be used as predictive tools does not prevent them from being useful in a policy-making process. Testing model sensitivity to deviations in input parameters, making slight variations in model structure or internal weighting schemes, and simultaneously using multiple models of the same phenomena can inform users as to the topography of the outcome space in which they are likely to find themselves. Such approaches not only reveal the limitations of the model or models, a form of sensitivity analysis, but also provide insights into the implications of unknowns and what is fundamentally knowable or unknowable about the likelihood of conflict. Although such uses do not produce predictive singularities, they can be employed to support real decisions by characterizing possible futures and identifying data and indicators that would suggest their emergence. While such results are weakly predictive, they have proven to offer powerful insights into problems that can only be analyzed by employing unvalidated or unvalidatable models.[61]

Robust Decisionmaking. As mentioned above, PCMT employment of social science models is intended to enable users to develop robust and adaptive policies. The shifting of decisionmaking criteria from optimality, the traditional focus of operations research analysis and modeling efforts, to robustness denotes a significant change in focus and the relationship between the known and unknown world. To fully appreciate this feature, it is important to acknowledge two features that confront policymakers working at the national level.

First, the specific features and character of the international system are fundamentally uncertain. Although policymakers may be guided by academic theories, generic knowledge, or rules of thumb, such models are heuristic and do not deserve unwavering intellectual or policy commitment. Indeed, this is specifically why PCMT incorporates a variety of different models and methods in its analytic suite to represent and broaden the sources of information and search space for solutions available to uses. The traditional decisionmaking paradigm of predict-then-act is not viable.[62]

Second, as decisionmakers move outside small groups of familiar colleagues and trusted peers into an increasingly broad and diverse environment, the criteria for policy fitness, or what constitutes an optimal, satisfactory, or unacceptable outcome, will increasingly diverge. Interpretations of the international system's structure, expectations of its behavior, and measures of effectiveness for policy will change based on the context of the debate and its participants.

Robust policies address situations where deep uncertainty reigns and traditional policy analysis methods break down. Policies or strategies are regarded as robust if they perform well across different models of the international system, even if they may not be ideal or optimal according to a particular model. Policy robustness against multiple

---

[61] See James A. Dewar, Steven C. Bankes, James S. Hodges, Thomas Lucas, Desmond K. Saunders-Newton, and Patrick Vye, *Credible Uses of the Distributed Interactive Simulation (DIS) System* (Santa Monica, CA: RAND, 1996), 24-35.

[62] See Robert J. Lempert, Steven W. Popper, and Steven C. Bankes, *Shaping the Next One Hundred Years* (Santa Monica, CA: RAND, 2003), 26-29.

models hedges against unknowns, while building in hooks or room for adaptation as time passes and more information becomes available.[63]

PCMT social vulnerability models reside within a framework called Computer Assisted Reasoning System (CARS). The CARS software performs two tasks. First, it performs broad parametric searches within each of the social vulnerability models, identifying model responsiveness to changes in input values and creating landscapes of outcomes. In practical terms, CARS allows users to evaluate model outputs over a range of values, creating a virtual laboratory where variations in data can be explored in relation to effects on the models individually and collectively. Ironically, as the ability to collect information improves, the probability of finding conflicting values for given variables is likely to increase. CARS enables users to perform multiple model runs, thereby allowing analysts to observe how outputs change vis-à-vis ranges of parameter values and enabling users to see how a gamut of values does or does not affect outcomes. In some cases large changes in input values have may negligible influence over model outputs; in other cases small changes in input values may have a large influence over model outputs; and in other instances certain outcomes may only be possible if a specific combination of values exist for a set of variables.

In addition CARS enables users to map various policy actions to model parameters, allowing for the exploration of actions and consequences within the models' outcome space. The CARS software includes a standard array of policy options wired to model parameters, and users can add policies and alter relations as desired. This allows users to rapidly examine how a generic suite of policies might perform while providing the flexibility for users to represent novel actions or alternative views of the effects of different actions. In combination, CARS allows for users to individually and collectively analyze the relationships between actions and consequences by providing a transparent, external environment to mediate policy debates.

Additionally, CARS allows for multiple, competing models to be used simultaneously. This allows for the outputs of different models to be compared and identifies the divergence and convergence of expectations across competing fitness criteria of multiple users. When combined with CARS' parametric search capabilities, the result is a broad array of futures based on multiple perspectives of intrastate conflict. While an individual model may not accurately forecast the future, the collective outputs of each model provide a range of outcomes or collection of scenarios, thereby enabling decisionmakers to contemplate what is possible, what to prepare for, and what is desirable.

Through CARS, users enter into an iterative cycle between choice making and analysis that simultaneously maximizes the robustness and minimizes the failure modes of their policies. By enabling users to map various strategies against changes in model input parameters, the effects of a given course of action can be simulated. By comparing how given policies or strategies fare across all models and ranges of initial conditions,

---

[63] See Robert J. Lempert, Steven W. Popper, and Steven C. Bankes, *Shaping the Next One Hundred Years* (Santa Monica, CA: RAND, 2003), 39-67. For a detailed account of robust adaptive policies see James A. Dewar, Carl H. Builder, William M. Hix, and Morlie H. Levin, *Assumption-Based Planning* (Santa Monica, CA: RAND, 1993); Paul K. Davis, "Uncertainty-Sensitive Planning," in Stuart E. Johnson, Martin C. Libicki, and Gregory F. Treverton, eds., *New Challenges, New Tools for Defense Decisionmaking* (Santa Monica, CA: RAND, 2003), 131-155.

specific actions can be systematically compared with their alternatives. Strategies that perform well across a variety of models and initial conditions, and allow for adaptation as new information becomes available are considered robust.

Interagency and International Collaboration

The character of political instability makes interagency and international planning, collaboration, and operations crucial. While military operations have a clear role to play in countering terrorist operations, drug trafficking, counter-proliferation, and deterrence, effectively dealing with the social, economic, and political conditions that motivate intra and interstate conflict requires a broader set of tools and engagement options.

While the PCMT Program's technological and analytic tools are state-of-the-art in their respective fields, the issue of interagency coordination and collaboration remains a compelling and timeless issue. The PCMT infrastructure and modeling suite is designed to facilitate interagency and even international collaboration to bring a more diverse set of resources and perspectives to bear on complex problems. It is important to recall that the PCMT database structure is designed to incorporate multiple datasets, contributions from non-traditional sources of information, and comments and discussion about its content. Furthermore, because the underlying PCMT data structure is ontology neutral and rapidly reconfigurable, any model can draw from the dataset, recoding the corpus as needed. Therefore, the analytic infrastructure is capable of soliciting multiple, competing perspectives on problems, much like the interagency process itself. The implications of the PCMT technical approach to data and analysis for collaboration are only just starting to emerge.

Traditionally, collaborative processes begin with a period of establishing definitions, terms of reference, norms of behavior, and other activities that bound the way the problems are represented and the solutions groups consider. The PCMT approach to collaboration is different. Rather than predefining the terms of reference and the nature of the debate, PCMT deliberately employs competing models to foster and facilitate analytic debate. This approach allows participants to use whatever models best represent their personal, institutional, professional, or other interests and concerns. Thus, diverse groups need not be forced to accept models or terms of reference they object to, encouraging their sustained interest and participation in collaborative planning and action. This pluralistic view of complex problems allows for greater and deeper exchanges between agencies.

Finally, the PCMT database and analytic architecture resides on a standard PC residing within a collaborative information environment. This information architecture enables users to communicate and share information in near real-time from distributed sites. As a result, analysts, consultants, advisors, and policymakers can communicate and share ideas on data, model results, policy proposals, etc.

Although PCMT is capable of transforming interagency planning and coordination, it will only do so if policymakers embrace its systems oriented process and operational concept. Their acceptance demands that PCMT technical infrastructure and software satisfy the look and feel of users regarding ease of use, visualization, speed, stability, etc. Likewise, more organizationally salient features must be considered, such as incorporating models or theories that user organizations consider credible in the analytic suite.

Recognizing that technology and concept development are an iterative process–linking research, development, and operators–PCMT has vigorously pursued relationships with, and feedback from, operators, policymakers, and analysts across the foreign policy community. This process of linking PCMT technological development and analytic and planning methodology with real-world decisionmakers has been facilitated by JFCOM.

The JFCOM Joint Interagency Coordination Group (JIACG) has served as the first test-bed for operational use and experimentation in support of the JFCOM Multi-National Experiment 3.[64] In addition, JFCOM conducted experiments on PCMT as a tool and process from February through July 2004. Experiments were conducted monthly, related to Central Asia. Afghanistan, Pakistan, Uzbekistan, and the Pakistan-Afghanistan border, each region receiving a different level of detail and focus based on operational and developmental requirements. For example, one experiment focused almost exclusively on the technological capabilities of the database software, while another aided in the production of an operational plan for use by the Joint Staff.

Whereas experiments customarily focus on fictional regions or crisis scenarios, PCMT provided participants with real data about the current state of the world. Because it is designed to provide users with situational awareness and the ability to prevent the outbreak of violence, experiments focused on evaluating whether PCMT could provide regional experts, analysts, and policymakers with new information about problems they deal with on a regular basis, and help them think about the complexities of the region and issues affecting political stability. As a result, PCMT users have been active participants in its development. With the completion of the first PCMT development cycle, JFCOM has published a concept of operations for its use in an operational environment.

---

[64] For additional information on the JIACG see http://www.jfcom.mil/about/fact_jiacg.htm.

PCMT and the Future

Since beginning in April 2003, PCMT has developed rapidly. At this time, individual technologies show promise, and early results demonstrate the utility of each PCMT subsystem as well as the uniqueness and power of the larger process that links them. At the conclusion of the eighteen-month proof of concept study, PCMT proved to be a successful research effort. However, these results do not fully reveal the future trajectory of the PCMT project.

PCMT experimentation efforts have identified several areas where additional development is required before an operational capability can be achieved. The PCMT software, i.e., CAST and Collier models, CARS, and database interface, need to be embedded into collaborative software tools, such as Information Workspace or Groove. In addition, the underlying file formats that transfer data between the database and the models are undergoing a change to improve transfer speed, processing efficiency, and reduce bandwidth demands. Other scheduled development activities include improving PCMT document coding methods through the use of neural networks, calibrating model responsiveness to policy actions to improve the functionality of the CARS environment, and creating a web-based architecture so that users can access the PCMT analytic suite through a portal. At this time, an initial operational capability could be available by fall of 2005 if the decision is made to transition the program from a proof-of-concept study into an operational capability. If operational, the PCMT Program's sustained development efforts include improving document analysis and automated scoring mechanisms, expanding the number of countries included in the database and analytic suite, adding additional social vulnerability and conflict prevention models, and developing additional data collection agents to cover a broader range of information sources.

In addition to the development activities that directly support the PCMT Program, DARPA is exploring a long-term research and development plan that will further enhance the functionality of PCMT. DARPA is developing a long-term research agenda focused on creating new social science models and methodologies for conflict analysis and the assessment of strategic effects. These modeling, simulation, and analytic methods include agent-based modeling and simulation, cellular automata, ARMIA statistical models, Ising models, and the incorporation of remote sensing data with social and economic behavioral models. Massive multiplayer online games (MMOGs) and other immersive environments are being developed to allow users to experience the effects of various strategies or model outcomes and better aid in the assessment of shaping strategies and conflict prevention efforts. Additionally, alternative means for automated hypothesis generation and testing and other forms of novel discovery are being examined for application to defense problems.[65]

The PCMT Program has demonstrated the potential to give policymakers and their staffs and advisors the ability to improve the effectiveness of conflict prevention

---

[65] Such novel discovery methods have been employed in the physical sciences for several years and are starting to be employed against social problems and adaptive systems. See Herbert A. Simon, *The Sciences of the Artificial* (Cambridge, MA: MIT Press, 2001), 105-110.

efforts. Continuation of the PCMT Program will give the United States a powerful new capability for preventing conflicts and shaping the international system, and will increase the ability of policymakers to foresee the consequences of their decisions and will transform planning and operations from focusing on winning wars to winning the peace.

www.ingramcontent.com/pod-product-compliance
Lightning Source LLC
Chambersburg PA
CBHW081410170526
45166CB00010B/3284